13.87

W9-BLT-954

Young Naturalist
Field Guides

Seashells, Crabs, and Sea Stars

by Christiane Kump Tibbitts
illustrations by Linda Garrow

Gareth Stevens Publishing
MILWAUKEE

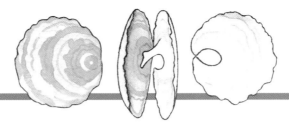

DEDICATION

For my parents, who first showed me the wonders of the ocean

ACKNOWLEDGMENTS

My heartfelt thanks go to Cristy Mittelstadt, Environmental Educator at the North Carolina Aquarium at Pine Knoll Shores for her suggestions and scientific review of the manuscript; to other Aquarium staff for their help; to Hugh Porter, Curator of Collections, University of North Carolina Institute of Marine Sciences, for answering questions; to Consie Powell and Marnie Brooks for editing and proofreading help; to my family — Dean, Alexandra, and Veronica — for their untiring support; to Diane Burns for cheering me on; to the Apex Public Library staff, Pamela Marino of the Cousteau Society, and Merrill Manke of the Monterey Bay Aquarium for graciously helping me obtain research materials.

For a free color catalog describing Gareth Stevens' list of high-quality books and multimedia programs, call 1-800-542-2595 (USA) or 1-800-461-9120 (Canada). Gareth Stevens Publishing's Fax: (414) 225-0377. See our catalog, too, on the World Wide Web: http://gsinc.com

Library of Congress Cataloging-in-Publication Data

Tibbitts, Christiane Kump, 1953-
    Seashells, crabs, and sea stars / by Christiane Kump Tibbitts ; illustrated by Linda Garrow.
       p. cm. — (Young naturalist field guides)
    Originally published: Minocqua, Wis. : NorthWord Press, © 1996, in series: Take-along guide. With new index.
    Includes bibliographical references and index.
    Summary: Explores the world of seashells, crabs, sea stars, sand dollars, and other things to be discovered at the seashore.
    ISBN 0-8368-2041-X (lib. bdg.)
    1. Mollusks—Juvenile literature. 2. Crabs—Juvenile literature. 3. Echinodermata—Juvenile literature. 4. Shells—Juvenile literature. [1. Mollusks. 2. Crabs. 3. Echinoderms. 4. Shells.] I. Garrow, Linda, ill. II. Title. III. Series.
QL405.2.T535 1998
592—dc21   97-41574

This North American edition first published in 1998 by
Gareth Stevens Publishing
1555 North RiverCenter Drive, Suite 201
Milwaukee, Wisconsin 53212 USA

Based on the book, *Seashells, Crabs and Sea Stars*, written by Christiane Kump Tibbitts, first published in the United States in 1996 by NorthWord Press, Inc., Minocqua, Wisconsin. © 1996 by Christiane Kump Tibbitts. Illustrations by Linda Garrow. Book design by Lisa Moore. Additional end matter © 1998 by Gareth Stevens, Inc.

Printed in Mexico

1 2 3 4 5 6 7 8 9 02 01 00 99 98

CONTENTS

Seashells, Crabs and Sea Stars

Metric Conversion Table

mm = millimeter = 0.0394 inch
cm = centimeter = 0.394 inch
m = meter = 3.281 feet

Note: Metric equivalents below are rounded off.

0.25 inch = 0.63 cm (6.3 mm)	5 feet = 1.5 m
1 inch = 2.5 cm (25 mm)	10 feet = 3 m
10 inches = 25 cm	15 feet = 4.6 m
12 inches (1 foot) = 30 cm	25 feet = 7.6 m
1 foot = 0.3 m	100 feet = 30 m
3 feet = 0.9 m	300 feet = 91 m

INTRODUCTION

As a wave sweeps the beach, tiny clams pop up. On a mud flat, a crab runs sideways and zips down a hole. In a rocky tide pool, a sea star moves slowly. Amazing animals live at the seashore. You can find them—with sharp eyes and patience.

Seashores can be sandy, muddy or rocky. They can be pounded by waves or protected in bays. A place where fresh water from rivers or streams mixes with salt water is called an estuary. Some places have salt marshes or swamps. Each kind of shore is home to many animals and plants.

The seashore is always changing. This *Young Naturalist Field Guide* and its activities will help you identify some amazing seashore creatures. You can use the ruler on the back of this book to measure the interesting things you find. You can bring a pencil and draw what you see in a notebook.

Always remember, if a shell feels full, or you can see something inside, the animal is alive and should not be disturbed.

Discover and have fun in the world of seashells, crabs, and sea stars!

SEASHELLS

Seashells are really the outer coverings of animals known as mollusks. Snails, clams and oysters are just a few examples. These mollusks make a chalky juice that hardens into a shell. The shell protects the mollusk's soft, squishy body. As the animal grows, it makes its shell larger. When mollusks die, their empty shells wash up on the beach for you to find.

Snails make shells that twist in a spiral. A snail glides around on a big, slimy foot. Many snails have a hard plate on the foot. When the snail hides in its shell, this plate closes the opening like a door.

A snail's head has two feelers and two eyes. The snail's tongue has rows of tiny, raspy teeth. It is called a radula.

Clams have two shells that open and close like a suitcase. The shells are joined at the top. A clam has a strong foot to burrow in sand or mud.

Some other kinds of mollusks live fastened to hard places. Most of these animals do not have heads or eyes.

LEWIS' MOON SNAIL

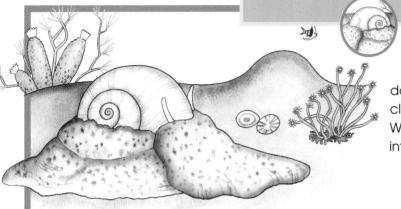

A moon snail eats about four clams a day. The snail wraps its huge foot around a clam and spreads strong acid on the shell. With its radula, the moon snail bores a hole into the shell and sucks out the clam.

This snail was named for Meriwether Lewis. He found it when he explored the Pacific Coast. Under a moving hump of sand, a moon snail plows along on its giant foot. The foot is so large it looks like it could never fit in the shell, but it does.

This moon snail's thick shell grows to about 3 inches long and 5 inches wide. The outside is yellow-white to pale brown. The inside is gray. The snail's foot is gray. It is flat and oval. It is about three times as long as the shell.

Lewis' moon snails can be found in sandy or muddy bays from British Columbia, Canada to California. Look in the sand near the low tide line for a wide track or a low hump.

The best times to explore are low tide or after storms.

FLAT PERIWINKLE

A flat periwinkle is about 1/2 inch high and just as wide. The shell is brown-gray with white spots. It often has many flat, worn places, but we don't know why. Inside the shell is dark brown with a white band on the bottom.

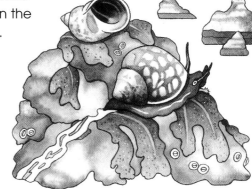

The flat periwinkle climbs rocks above the high tide line, higher than any other marine mollusk on the Pacific Coast. Over many years, flat periwinkles left the sea and slowly learned to breathe air. Flat periwinkles live out of the water most of the time. They can still breathe under water, too.

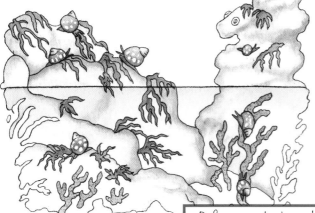

Flat periwinkles slide slowly across rocks on a trail of slime. This comes out of a slit in the snail's foot. A periwinkle moves like a skater on ice. The snail glides on one side of its foot, then the other. Flat periwinkles wander, scraping seaweed off rocks with their radulas. These snails may be found feeding even during low tide.

Flat periwinkles can be found from Washington to California.

Before exploring, check the newspaper for low and high tide times— they change each day.

ATLANTIC OYSTER DRILL

Atlantic oyster drills like to eat oysters. First, the oyster drill crawls on top of the oyster shell and feels around for the best place to drill. Acid comes out of the snail's foot and eats into the shell. Then the oyster drill scrapes with its radula. For three days, the oyster drill keeps spreading acid and scraping. Finally there is a small, straight hole through the shell. The oyster drill uses its long tongue to eat the oyster.

The inside of the shell is purple, white, yellow or brown. The snail is creamy white.

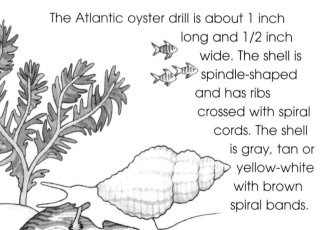

The Atlantic oyster drill is about 1 inch long and 1/2 inch wide. The shell is spindle-shaped and has ribs crossed with spiral cords. The shell is gray, tan or yellow-white with brown spiral bands.

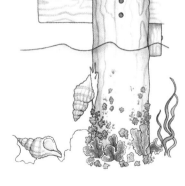

Atlantic oyster drills live on oyster beds, rocks and docks in the saltier parts of estuaries and on bay and ocean shores. You can find them from Nova Scotia, Canada to Florida, and from Washington to California.

Wear old sneakers to protect your feet from sharp objects.

LIGHTNING WHELK

A small lightning whelk fits in your hand. But a big one is more than a handful. The shell grows about 3 to 16 inches long. The snail's foot is black.

A snail's shell grows in an ever-larger spiral. You can see this at the top of a shell. On most snail shells the spiral twists to the right. On some, it twists to the left. The lightning whelk is a left-handed shell.

A lightning whelk likes to eat hard-shelled clams. This whelk often has a battered shell from banging against clams to crack them open. Then the whelk scrapes the clam out with its radula.

The lightning whelk's shell is thick. It is wide on one end, and comes to a point at the other end. The shell is yellow to gray-white. It has thin red-brown "lightning" streaks from end to end. The inside is white.

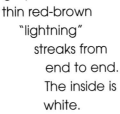

Lightning whelks live along sandy beaches from New Jersey to Florida and Texas. They are also found in the saltier parts of estuaries.

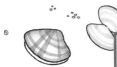

Always put rocks back the way you found them.

CAYENNE KEYHOLE LIMPET

A cayenne keyhole limpet is oval. It is 1/2 to 2 inches long. The shell is ribbed and white, cream, gray or pink. The inside is shiny and blue-gray or white. The snail is creamy white.

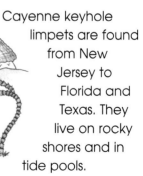

Keyhole limpets are most active at night during high tide. A limpet wanders over rocks, scraping off seaweed with its radula. At low tide, each limpet returns to a home spot it has chosen.

A cayenne (pronounced "KI-EN") keyhole limpet's shell looks like a small volcano. The hole on top is shaped like a keyhole. This gives this limpet its name. A keyhole limpet pulls water in under its shell to breathe. Water and wastes go out the hole on top.

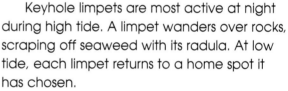

Cayenne keyhole limpets are found from New Jersey to Florida and Texas. They live on rocky shores and in tide pools.

The snail's strong foot clamps down like a suction cup. Its cone-shaped shell is pressed down even tighter when heavy waves hit.

Bring a small shovel and plastic bags for collecting things.

BLACK KATY CHITON

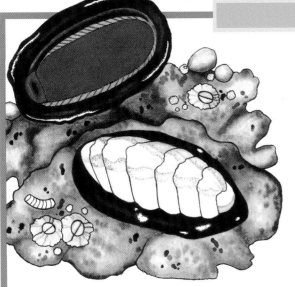

A chiton can hang onto a rock like a suction cup by raising its body up inside its girdle. If it loses its grip, a chiton curls up in a little ball, safe inside its armor.

The black Katy chiton (pronounced "KI-TON") is 1-1/2 to 3 inches long. It is covered by a row of eight gray or blue-white plates of shell. Each plate is curved and shaped like a boomerang. The plates are part of the soft animal's back. They overlap each other, forming an oval shield. Around the edge is a shiny black, stretchy band of muscle called the girdle.

Under all this armor is a very simple mollusk. Most of it is a large, strong red foot. At one end of the body is a mouth with a radula.

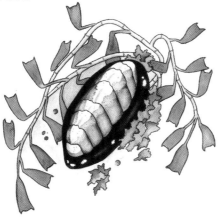

A black Katy chiton is a dull color like rock. This helps the animal hide from enemies. Black Katy chitons move around in the daytime, chewing seaweed off rocks. You can find them on rocky shores from Alaska to California.

A snorkeling mask placed on top of the water in a tide pool helps you see into it.

EASTERN OYSTER

The eastern oyster grows to be about 2 to 10 inches long. The shell is oval, with sharp edges. It is thick and wrinkled. The outside is gray. The inside is white with purple. The animal inside is gray.

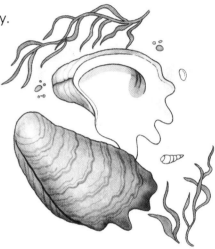

When oysters are young, they find a hard spot to stick to and stay there for life. Other oysters settle nearby. They grow. New oysters settle on top of them. In time, an oyster city of many layers rises up.

No two oysters are alike. A growing oyster is shaped by what it lays on. Its two shells are uneven. The bottom shell is cupped. The top one is flat. An oyster opens its shell to strain tiny plants and animals from water. At low tide, oysters close.

Oysters like a mix of salty and fresh water. They live in shallow water in estuaries, salt marshes and mangrove swamps. You can find eastern oysters along the shore from the Gulf of the St. Lawrence River to the Gulf of Mexico.

Sunscreen, layered clothes and a sun hat keep you comfortable.

COON OYSTER

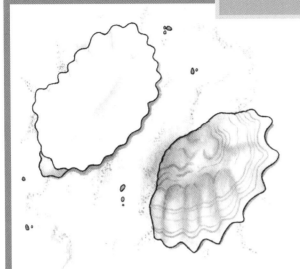

A coon oyster has hooks to hold onto things. Its bottom shell grows spines like little fingers. Coon oysters grip the arched roots of red mangrove trees. This way, the oysters stay above the mud. At low tide, they are uncovered by water. Early Spanish explorers thought these oysters climbed trees!

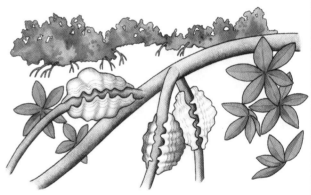

Coon oysters strain the food they eat from the water. They are food, too. At low tide at night raccoons eat them. That is how coon oysters got their name.

A coon oyster is about 1 to 3 inches long. The two shells are roughly oval. They are yellow-white, rosy or brown. The inside is green or white and purple. Big, uneven ridges make a zigzag around the shell's edge. Sometimes coon oysters cling to each other in a bunch. They also live on brush in shallow water. Coon oysters are found from North Carolina to Florida.

SPECIAL WARNING Watch out for alligators and snakes in mangrove swamps, and go with a guide.

COMMON JINGLE SHELL

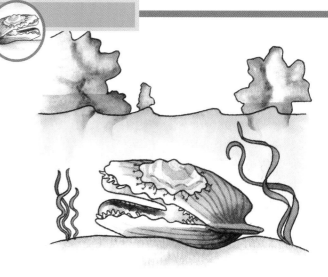

This round shell is so thin you can see through it. Jingle shells look like toenails. Sometimes jingle shells are called "mermaid's toenails."

The common jingle shell is pearly white, gold, apricot, silvery gray or black. It is about 1 to 2 inches long. Jingle shells come in pairs. The two shells are not the same. The top one is cupped and the bottom one is flat. The bottom shell has a hole. Through this hole, the animal sticks itself to a hard place.

These threads harden in seawater. The jingle is stuck for life in the place it has chosen. As a jingle grows, its shell becomes shaped to fit its home.

Jingle shells live stuck to rocks, docks and other shells like oysters. Sometimes jingles also stick to boats or horseshoe crabs. A jingle fastens itself by spinning sticky threads with its foot.

Common jingle shells can be found from Nova Scotia, Canada to Florida and Texas. You can find these shells on sandy or rocky beaches.

Be careful climbing on rocks— they can be slippery.

COQUINA CLAM

Coquinas move with the tides. From time to time, all the clams pop out of the sand at once. They ride a wave to a new place on the beach and dig in.

These clams are about the size of dried beans. Coquinas (pronounced "KO-KE-NA") live jam-packed just beneath the sand. There could be 1,000 of them under your feet!

Coquinas are 1/2 to 1 inch long. They come in a rainbow of colors. The outside is white with bands or sunrays of pink, purple, blue, orange or yellow. The inside is white with yellow, purple or pink.

Each coquina digs by stretching its pointed foot down in the sand. Then the clam pulls in its smooth shell. The triangle shape helps the shell slide in easily.

Coquina clams live where the waves pound sandy beaches from New York to Florida and Texas. You may find bunches of open coquina shells still joined together. Some people call these butterfly shells.

SPECIAL WARNING

Stay away from jellyfish—their stingers are poisonous.

NORTHERN QUAHOG

Northern quahogs live buried just beneath the sand and mud. When you step close, a quahog squirts water as it digs down to get away. This clam can dig fast with its strong foot.

Long ago, Native Americans admired these shells and made them into beads. This was done by drilling and polishing pieces of shell. Belts of these beads were valuable. Some people call the quahog (pronounced "KWA-HOG") the hard-shelled clam because it has a thick shell.

The northern quahog is oval. It is about 3 to 6 inches long. The two shells look the same. Each is gray-yellow outside, often with brown. Lines run across the shell. The inside is white and purple.

Quahogs strain tiny plants and animals from seawater. This also helps the quahog breathe. An adult quahog pumps through more than 15 quarts of water an hour.

Look for northern quahogs on tidal flats in bays and along ocean shores. These clams can be found from Quebec to Florida and Texas, and in California.

Protect the seashore. Don't take live animals or plants.

CALICO SCALLOP

A scallop has over 30 bright blue eyes. They peek out around the edge of its shell. The eyes see only shadows, but help the scallop escape from enemies.

The calico scallop has two round, cupped shells. One is flatter than the other. Each shell is about 1 to 3 inches long. Ridges on the shell make a "scallop" pattern at the edge. Calico scallops are spotted or mottled. They may be any mix of white, yellow, brown, orange, red or purple. The inside is white.

Most mollusks move slowly, or not at all. But scallops can swim. A scallop swims by clapping its shells together. Water squirts out one side, pushing the scallop the other way. A scallop can jet 3 feet at a time! It swims to keep from getting buried in mud.

Calico scallops lie on their flat side on sand or mud in shallow to deep water. You can find their shells on sandy beaches from North Carolina to Florida and Texas.

Some places are muddy, so be sure to wear boots.

HEART COCKLE

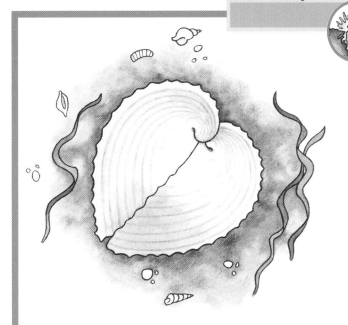

Cockles burrow in sand or mud, but not very deep. When an enemy comes near, a cockle leaps away. It bends its strong foot and pushes off the sand. A cockle can jump several inches at a time.

A cockle has two shells that look the same. Each shell is oval to almost round. It is deeply cupped with thick ridges. The shell edge looks ruffled. From the side, a cockle looks like a heart.

The heart cockle is 2 to 5-1/2 inches long. The shell's outside is gray with a thin, flaky yellow-orange or yellow-brown covering. The inside is yellow-white. The cockle's foot is bright yellow and shaped like a hatchet.

Heart cockles live in quiet bays and estuaries. They also live along open shores. Cockles can be found from the low-tide line to deep water. Look for holes in the sand or mud. Sometimes a cockle lies above the sand. But it can burrow in seconds. Heart cockles are found from Alaska to California.

Don't leave litter on the shore—
it can hurt or kill animals.

JINGLE-JANGLE WIND CHIME

Seashells are not only beautiful, but they can also make music in the wind. Many seashells have holes in them. The bottom parts of jingle shells grow with holes in them. But snails like oyster drills and moon snails bore the holes in other shells. These are shells you can use to make this wind chime.

WHAT YOU NEED

▼

- 12 feet of string

- a washable marker

- a 12-inch ruler or a yardstick

- scissors

- lots of seashells, each with a hole in it

- a piece of driftwood or a thick stick

An adult can help you hang your wind chime outdoors where everyone can admire it and listen to the music!

WHAT TO DO

▼

1 Cut 4 pieces of string each 3 feet long.

2 Mark 3 strings 8 inches from one end.

3 Put one string through the hole of one shell. Slide it to the 8-inch mark. Tie a knot. Below it, tie on another shell close to the first one. Tie on shells the rest of the way down the string. Do the same for the other 2 marked strings.

4 Tie one end of each string of shells to the piece of wood.

5 Push the strings close together so the shells touch.

6 Tie the last piece of string to the piece of wood to make a loop for hanging.

SEASHORE RUBBINGS

Many things are fun to touch at the seashore. Bumpy shells, smooth pebbles, twisty driftwood and seaweed are a few things you can collect to do this fun activity.

WHAT YOU NEED

- some objects to rub over
- tracing paper
- transparent tape
- crayons (with the paper covers off)
- scissors

WHAT TO DO

1 Put an object on a flat place where you have room to work. If you can, tape down the object.

2 With one hand, hold the paper down tightly over the object.

3 With the other hand, rub a crayon on the paper over the object. Use the side of the crayon and rub hard. Try to make all the edges look clear and sharp.

4 Try many different kinds of things. You can have just one object on the paper or you can have several.

You can make a picture with rubbings. You can create bookmarks and letter paper. Or you can use paper with rubbings all over it as gift wrap.

CRABS

Crabs, like seashells, have outer shells to protect their bodies. A crab's shell has many plates joined together. The shell does not get bigger as the crab starts to grow. When its shell becomes tight, the crab grows a new one under it. Then, the old shell splits and the crab wriggles out of it. After three days its new, soft shell becomes hard.

A crab has two eyes on stalks to see in all directions. Between the eyes are two pairs of feelers. Crabs have 5 pairs of legs. The front pair has claws used to grab food and for fighting. The other legs are used for walking or swimming. A crab can grow new legs to replace the ones it loses!

Some crabs live in the sea. Others live on land. Crabs eat seaweed and small sea animals.

Mole crabs and barnacles are not true crabs, but are related more to them than to any other animal. Mole crabs don't have claws. Barnacles are the only crab cousins that live attached to the same place for life. Horseshoe crabs are actually more closely related to spiders!

GHOST CRAB

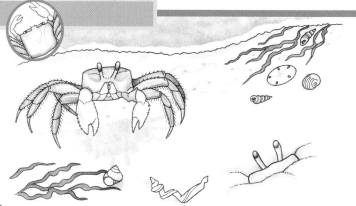

A ghost crab can run fast—up to five feet per second! This crab runs on the tips of its eight long legs. It can run forward, backward or sideways. It can turn and switch legs while running in the same direction.

A ghost crab burrows in damp sand. The crab warns enemies away by making a creaking sound. It does this by rubbing its two claws together. Before leaving its burrow, the crab peeks out. Its eyes are on long stalks.

At night, groups of ghost crabs come out. They hunt small animals like mole crabs and coquina clams. Ghost crabs also eat dead plants and animals washed on the beach.

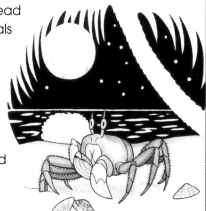

These crabs live on quiet, sandy beaches from Rhode Island to Florida and Texas.

Ghost crabs are pale gray or yellow like sand. Its body is 2 inches wide, but its legs make it look bigger. The legs are about 4 to 6 inches long.

SPECIAL WARNING

Watch out for rising tides!

LONG-CLAWED HERMIT CRAB

When a hermit crab grows too large for its shell, the crab looks for a bigger shell. It lifts and tries on many empty shells. Finally it slips out of its old shell into a new one.

This crab has a hard shell only on the front part of its body. Its back end is soft, and easy to attack. So a hermit crab wears an empty snail shell like a suit of armor. Its claws and front legs stick out of the shell. One claw is bigger than the other. When the crab hides inside its shell, the bigger claw blocks the opening.

The long-clawed hermit crab is about 1/2 inch long and 1/2 inch wide. It is tan, lavender-gray or green-white. It has a tan-gray stripe on its big claw.

The long-clawed hermit crab lives in shallow to deep water on sand, mud, rock and in seaweeds. It lives along ocean or bay shores from Nova Scotia, Canada to Florida and Texas.

Salt marshes are fragile— explore carefully.

SAND FIDDLER CRAB

A male sand fiddler crab has one claw that is much bigger than the other. The big claw is longer than its whole body! It looks as if it is holding a fiddle, or violin.

The male sand fiddler shows off with his big claw. He holds it out, rising on tiptoe. He bends the claw and bows. He does this over and over, bobbing up and down. He often taps his big claw on the ground. In summer hundreds of male fiddlers do this dance. Their eye stalks are raised high. Each male hopes a female will notice him.

This crab's body is about 1 inch long and 1 inch wide. The top of the crab is lavender. Brown, gray or black spots are on the sides. The male's big claw is blue, lavender or red-brown. The female fiddler is darker and duller.

Sand fiddler crabs are found from Massachusetts to Florida and Texas. They live in sandy areas of salt marshes nearest the sea. These crabs also live on sandy bay beaches.

Use insect repellent to protect yourself from biting flies.

HORSESHOE CRAB

A horseshoe crab has six pairs of legs. Between the last three pairs is its mouth. The first two legs have small claws. The other legs have bristly spines at the top. These spines grind food and move it to the mouth as the crab walks. The last two legs end with "brushes." A horseshoe crab uses them like ski poles to push itself along.

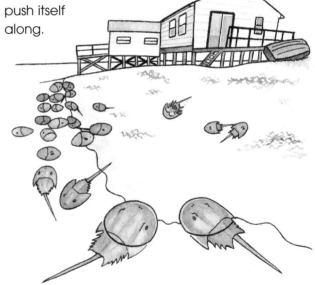

Horseshoe crabs lived on earth even before dinosaurs!

A horseshoe crab has one large eye on each side of its shell and two smaller eyes in front. It can see only light and dark. A horseshoe crab is 12 inches wide and 24 inches long. Its name comes from its shape.

This crab's long, pointed tail is harmless. If the crab gets turned over, it uses its tail to turn itself right side up.

These crabs live on sand or mud from shallow to deep water. In spring and summer, they come on beaches to lay eggs. You can find horseshoe crabs from Maine to Florida and Texas.

Don't pick up horseshoe crabs by the tail—it hurts them.

BAY BARNACLE

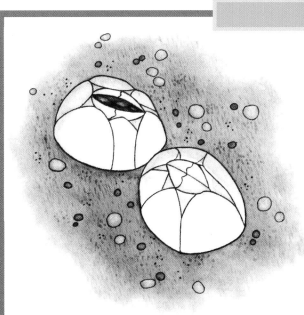

A bay barnacle is white. It is 1/4 inch high and 1/2 inch wide. The shell has sharp edges.

When young, a barnacle glues its head to a rock, dock, shell or boat. It might even live on an animal like a horseshoe crab! A barnacle grows a hard white shell for a home. Overlapping shell plates form walls in the shape of a circle. At the top, four plates make a door. It opens like the petals of a flower. When closed, the door seals in water so the barnacle won't dry out.

Inside its shell, a barnacle is upside down. When it opens its doors, the barnacle sticks out six pairs of hairy legs. They sweep tiny floating plants and animals into the barnacle's mouth. Barnacles crowd together in their crusty homes. Sometimes younger barnacles live on top of older ones.

You can find bay barnacles in estuaries from Oregon to California, and from Nova Scotia, Canada to Florida and Texas.

 SPECIAL WARNING *Stay away from coral reefs.*

STRIPED SAND SOMETHING-OR-OTHER

Sand is washed on the beach one layer at a time. You can see this if you dig a deep hole. Sand layers look like stripes. Sand can slowly harden into stone. Sandstone has stripes, too.

WHAT YOU NEED

- a wide, clear glass jar with its lid (like a large peanut butter jar)

- sand — a bit more than the jar will hold

- food coloring in different colors

- 1/4 cup measure

- clean, wide plastic cups — 1 cup for each color of sand

- a tall plastic drinking cup that fits easily inside the jar

- liquid glue

WHAT TO DO

▼

Your jar of striped sand can be a paperweight. Or you can just admire it on your dresser.

1 Put 1/4 cup of sand in a plastic cup. Add 8 to 10 drops of one food coloring. With a clean spoon, stir the sand and coloring. Make sure they are well mixed. Let the sand dry.

2 Repeat step 1 for each color of sand.

3 Pour colored sand from one of the cups into the jar. Using the bottom of the plastic drinking cup, press the sand down firmly. Add a different color of sand. Press it down firmly.

4 Add more layers of colored sand. Press each layer down before adding the next one. Fill the jar until sand is even with the top.

5 Put glue on the inside edge of the lid. Twist the lid on the jar. Let the glue dry.

Sea stars and their relatives have spiny skin. They each have a body that can be divided into five parts. Sea stars, brittle stars and sea cucumbers can grow new body parts to replace lost ones. Urchins and sand dollars can repair holes in their skeletons.

All these animals have hard plates inside their skin. In urchins and sand dollars, the plates are joined in a rigid skeleton.

To move, most members of the sea star family pump seawater through their bodies into tube feet. When water is pumped out, the tube feet grip like suction cups and the animal pulls itself forward. A sea cucumber pumps body fluid instead of seawater into its tube feet.

Most spiny-skinned animals have no eyes. But sea stars have a red eyespot on the tip of each arm. They can see light and dark.

Sea vases belong to a group of animals called sea squirts. Sea anemones live attached to hard places and have poisonous stingers.

OCHRE SEA STAR

The ochre sea star likes to eat mussels. First it climbs onto a mussel. The tube feet stick to the mussel's shell. With its arms, it pulls the mussel open. Then the sea star pushes its stomach out through its mouth into the mussel. After eating it, the sea star pulls its stomach back inside.

Ochre (pronounced "O-KER") sea stars can climb straight up a rock wall. They can cling to rocks in crashing waves and open mussels.

This sea star is 20 inches across. It has 5 thick arms. On the underside are many little tube feet. Its mouth is in the center. Ochre is a yellow or reddish-yellow color. Ochre sea stars are yellow, orange, brown, red or purple and have short white spines.

The ochre sea star is found from Alaska to California. It lives on rocky shores pounded by waves. It is often found on mussel beds and in tide pools.

SPECIAL WARNING

Don't go out on rocks when waves are crashing.

DAISY BRITTLE STAR

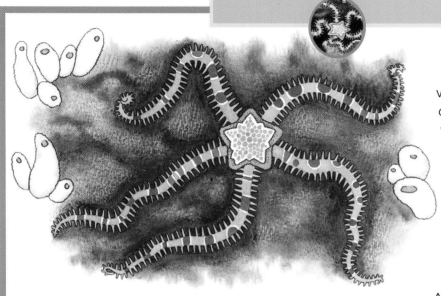

When scared, a brittle star waves its thin arms. It crawls away fast. It pulls itself along with one or two arms. The rest trail behind or push off the bottom. If something snags an arm, it breaks off easily. But the brittle star just grows a new one. This happens so often that their arms are usually different sizes.

A brittle star's arms are handy for catching small crabs or other food. It daintily stuffs them into its mouth underneath.

The daisy brittle star has five long, skinny arms. The arms are about 4 inches long with fine, blunt spines. The brittle star's cookie-shaped body is about 1 inch across. Daisy brittle stars are spotted or striped. They can be red, orange, pink, yellow, white, blue, green, tan, brown, gray and black. They look like bright flowers. That's probably how they got their name.

Daisy brittle stars hide under rocks in tide pools. They are found from Canada to Massachusetts, and from Alaska to California.

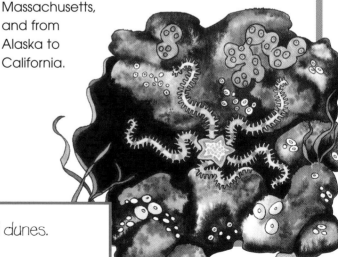

Stay off sand dunes.

PURPLE SEA URCHIN

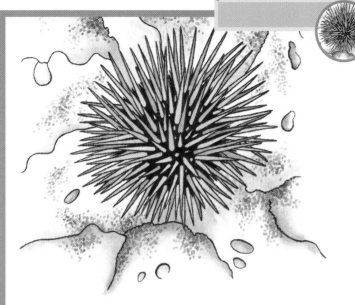

An urchin's mouth is underneath. It has five teeth that open and close like a beak. The purple sea urchin catches and eats floating seaweed with its tube feet. It also chews seaweed off rocks.

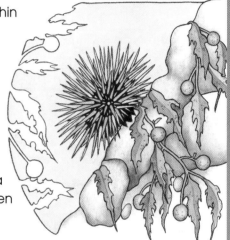

A sea urchin hides from enemies. It holds bits of shell and seaweed over itself with its tube feet. Or it may hide in a crack between rocks.

The purple sea urchin has spines all over. Among the spines are its tube feet. It looks like a porcupine. When an urchin walks with its tube feet, the spines act as stilts. It can move quickly. If an urchin turns over, its spines help it turn right side up.

A purple sea urchin is about 4 inches wide and 2 inches high. Its sharp spines are bright purple. Some kinds of urchins have poisonous spines, but not the purple urchin.

Groups of urchins live on rocks in shallow to deep water. They also live in tide pools. Purple sea urchins are found from British Columbia, Canada to California.

Don't pull off attached creatures. They will die.

ECCENTRIC SAND DOLLAR

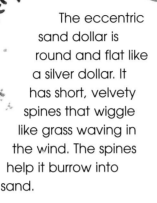

The eccentric sand dollar is round and flat like a silver dollar. It has short, velvety spines that wiggle like grass waving in the wind. The spines help it burrow into sand.

The eccentric sand dollar is about 3 inches across and 1/2 inch thick. Its spines are lavender-gray, red-brown, brown or purple-black. On top, it has a five-point star with tiny holes around it. Tube feet stick out of the holes. On its underside are more tube feet.

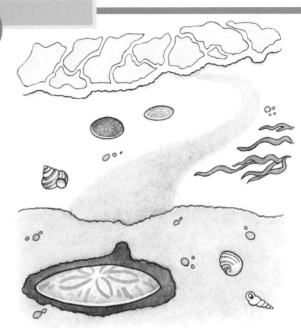

Thin grooves help guide food to its mouth in the center. The mouth has 5 small teeth that open and close like a beak. Sand dollars eat tiny plants and animals sifted from the sand. If you shake a sand dollar's skeleton, you'll hear the teeth rattling inside.

Eccentric sand dollars are found along the shoreline from Alaska to California. Groups of them burrow in the sand in shallow to deep water. Look in the sand for a flat track like a ribbon—a sand dollar may be buried where the track ends.

Bring fresh drinking water with you.

RED SEA CUCUMBER

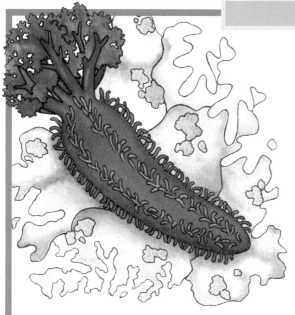

The red sea cucumber has two rows of tube feet on top. Three rows are underneath. It moves on these tube feet. Its muscles help push it slowly along.

When attacked, a sea cucumber squirts out most of its insides! While the attacker eats these, the sea cucumber escapes. New insides soon grow back.

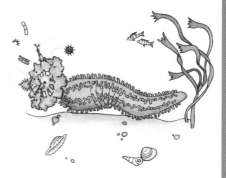

This creature's body is shaped like a cucumber. The skin is smooth but tough.

The red sea cucumber is about 10 inches long and 1 inch wide. It can be red, orange, pink or purple. At one end of its body is its mouth, with ten frilly orange feelers around it. Tiny floating plants and animals stick to the feelers. The sea cucumber pulls the feelers one by one into its mouth to eat the food.

This sea cucumber hides between rocks with its body curved so each end sticks out. They live in shallow water near the low-tide line. Red sea cucumbers can be found from Alaska to California.

Bring a flashlight for a night-time walk with an adult.

GIANT GREEN SEA ANEMONE

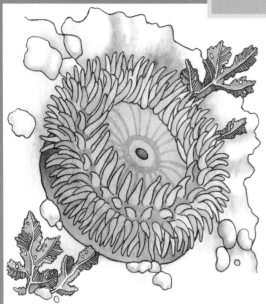

This sea anemone lives glued to rock. Sometimes dinner doesn't swim by for a long time. So, the anemone melts its glue and slowly glides to someplace new.

The giant green anemone can grow to be 12 inches high and 10 inches wide. The anemone's stalk is green-brown. The feelers are green, blue-green or white. The center is green, gray-green or blue-green.

Though it looks like a flower, the giant green sea anemone (pronounced "A-NEM-MEN-EE") is really an animal. Its body is a stalk, but petals aren't on top. Those are feelers with poisonous stingers. The feelers surround the anemone's mouth. When food swims near, the anemone stings it. Then the feelers pull the food down inside the anemone's mouth.

The giant green sea anemone lives along ocean and bay shores from Alaska to California. It is found on rocks and seawalls from above the low-tide line to deep water. It also lives in tide pools. When uncovered by the tide, giant green sea anemones close up to hold water inside.

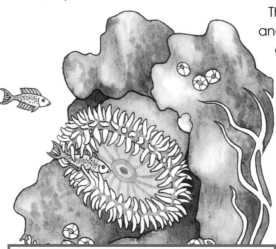

Explore safely— go with a partner.

SEA VASE

A sea vase is about 6 inches high and 1 inch wide. It is pale yellow or green. You can see a sea vase's organs inside. On top, a sea vase has two short tubes close together. They are edged with yellow. A sea vase pumps water in one tube and out the other. It strains tiny plants and animals from the water to eat.

If you touch a sea vase, look out! It will squirt water at you. That's why this animal is also called a sea squirt. The sea vase has a body like a bag. It feels like plastic. Sand and other things may stick to its body. Seaweed often grows on it. On the bottom are bumps like roots. A young sea vase fastens its roots to a hard place.

Sea vases live on rocks, docks and boats in quiet bays. These animals live in shallow to deep water. You can find them from the Arctic to Rhode Island, and from Alaska to California.

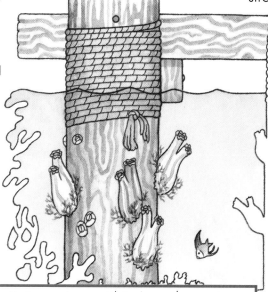

Bring along this book and a magnifying glass.

SEASHORE TREASURE CHEST

You've found interesting things at the seashore. Now you're ready to set up your collection to show your friends.

WHAT YOU NEED
▼

- empty egg cartons or shoe boxes with lids

- cotton balls

- little pieces of paper for name tags

- pencil or pen

- scissors

- liquid glue

WHAT TO DO
▼

First, sort your seashore treasures by shape. Then follow these steps for each one:

1 Rinse off the seashell. If needed, carefully scrub it with an old, soft toothbrush.

2 While the seashell is drying, find out what it's called.

3 Write the name on a piece of paper. You can also put the date and place you found it.

4 Put cotton in an egg carton hole or on the bottom of a shoe box. Lay your object on top of the cotton.

5 Put a dot of glue on the back of the paper. Press it on the cotton beside the object.

Sea Star Sand Dollar

Now your treasures won't get broken. And they're ready to show to your family and friends.

MAKE YOUR OWN SEASHELLS

You can make your own seashells to keep and decorate. Or you can give them away to friends so they can have "shells" of their own!

WHAT YOU NEED ▼

- some seashells (shiny ones with bumps and ridges work best)
- modeling dough or clay that doesn't get hard
- a clean plastic tub (like a margarine tub)
- an empty shoe box
- plaster of Paris and a stirring stick
- watercolors, paint, markers or crayons—anything you want to color them with

WHAT TO DO

1 Choose the shell you want to make first.

2 Knead the clay until it's soft.

3 Shape it into a smooth lump.

4 Press the bottom of the lump of clay into the box (to catch spills).

5 Place the shell on the clay like a cup (round side down).

6 Push the shell into the clay. Do not push all the way through to the box.

7 Carefully take the shell out of the clay.

8 Gently mix the plaster in the large plastic tub.

9 Slowly pour the wet plaster into the hole your shell made and fill it up.

10 Let the plaster dry at least 1 hour.

11 Carefully take the clay and plaster out of the box.

12 Pull the clay off the plaster.

13 Let the plaster dry at least 24 more hours.

14 Decorate your shell.

For More Information

MORE BOOKS TO READ

Crustaceans: Armored Omnivores. Secrets of the Animal World series. Andreu Llamas (Gareth Stevens)

Mussels: Hard-Shelled Mollusks. Secrets of the Animal World series. Andreu Llamas (Gareth Stevens)

Ocean Life. Under the Microscope series. Casey Horton (Gareth Stevens)

Sea Stars. Jason Cooper (Rourke Publications)

Seashells. R. Tucker Abbot (Thunder Bay)

VIDEOS

The Crab. (Barr Media Group)

Crustaceans. (Encyclopædia Britannica Educational Corporation)

Marine Invertebrates. (Encyclopædia Britannica Educational Corporation)

WEB SITES

www.wh.whoi.edu/homepage/faq.html
www.bev.net/education/SeaWorld/coral_reefs/introcr.html

INDEX